Rajesh Kumar Mahato

Morphometric Analysis of Sapahi River Basin

GRIN Verlag

Bibliografische Information der Deutschen Nationalbibliothek:

Die Deutsche Bibliothek verzeichnet diese Publikation in der Deutschen National-bibliografie; detaillierte bibliografische Daten sind im Internet über http://dnb.d-nb.de/ abrufbar.

Imprint:

Copyright © 2009 GRIN Verlag GmbH
Druck und Bindung: Books on Demand GmbH, Norderstedt Germany
ISBN: 978-3-656-32814-8

MORPHOMETRIC ANALYSIS OF SAPAHI RIVER BASIN

Rajesh Kumar Mahato

Research Scholar, University Department of Geography,

Ranchi University, Ranchi-8, Jharkhand, India

A B S T R A C T

Sapahi river is a very small tributary of river Swarnrekha river system. It's just 15 km far away from main capital and functional region of Ranchi. In the lower portion of river Sapahi, Rukka dam is major water reservoir for water supply in Ranchi capital. The case study considered as try to prove river's natural behavior. Thus it is a macro level field based research, there is very less possibility to acquire micro level field data. River rises from the up land of the northwest, area characterized by rocky explore and mainly forest with Sal. In monsoon season, river gives huge water mass, where as in winter it converts in a small stream. River deposits alluvial materials each year in maximum fluvial year. River morphometry and their related characteristics such as channel morphometry, drainage density and basin configuration are try to represent through different data and real environment calculation.

This paper is considered to explore some specific characteristics of very small river system, that how Sapahi river system behaving with their surface configuration.

KEYWORDS: *Bifurcation ratio, Lithological, Hypsometric Curves, Channel Geometry, Clinographic, Thalwag.*

RESEARCH METHODOLOGY:

A three-tier approach comprising image interpretation, intensive field survey and laboratory investigation as cartography and printing adopted. Image interpretation comprises data derivation from map analysis, map work, computation from topographical maps and satellite imageries pertaining to deferent component of settlement. Research also used Geospatial techniques for the landform analysis and various geographic changes.

In the last for empirical analysis on the spot verification on examination of fact dealing in surface configuration and river behavior taken as consideration through field observation and Photography.

HYPOTHESIS:

The main hypothesis is following types which I considered for testing:

- ✓ Charging and Discharge of Sapahi river basin differs in different seasons.
- ✓ Degradational behavior of Sapahi river basin is based on the Lithological characteristics.
- ✓ Basin morphometry depends on the configuration of surface and runoff characteristics.

INTRODUCTION

Sapahi River is a very small drainage unit; 13.8 km long and almost few km in width. It is a tributary of Subarnrekha river which rises from the up land of the northwest on the area characterized with rocky explore and mainly forest with Sal. Jirabar is a major tributary of Sapahi in initial stage, which joins near Jirabar village. Sapahi river joins with Subarnrekha (Rukka Dam) near Rukka village. The pattern of the drainage looks like dendritic. The general slope of the basin towards north-northwest to south-south-eastward. Through empirical observation, fully Granite and Gneisses characterize the river; underlying rocks are hard. From upper course to lower valley the river serves number of villages, which settled, beside the bank. During rainy season it appears with huge amount of water.

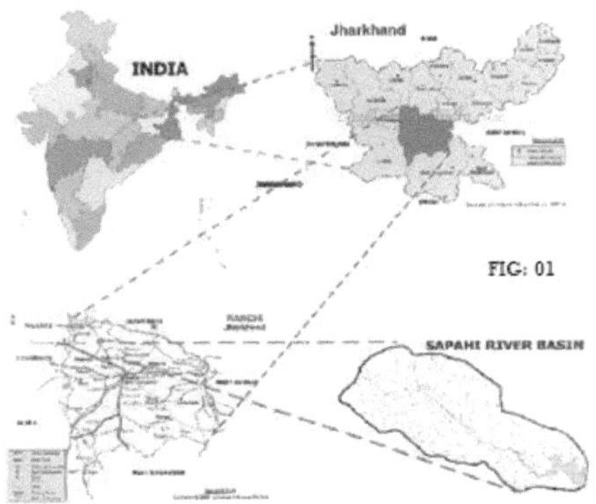

1.1 LOCATION OF THE STUDY AREA

The study area is located in Chottanagpur plateau region just 15 km far away from Ranchi main capital region. Problem selected i.e. the status of Sapahi river and their feeding capacity for living population in the valley, with special reference to resource evaluation – a macro-level study of the river Sapahi. Sapahi river basin is lying over Chottanagpur plateau of Jharkhand, especially say Ranchi Plateau. It has formation of consultation of molten magma since the Achaean period. The whole plateau is dominated by the rocks of granite and gneiss.

1.2 MORPHOMETRIC ANALYSIS

Morphometric concept refers to the real configuration of the earth surface. How the relief is is showing by our eyes. What are their 3D feature and their other things (Physical and Cultural)? The basin morphometry includes the analysis of the characteristics of linear, areal and relief aspects of fluvial originated drainage basin.

1.3 LINEAR ASPECTS OF THE BASIN

It includes channel patterns of the drainage network wherein the topological characteristics of the stream segments in terms of open links of the network system (streams) are analyzed.

1.3.1 STREAM ORDERING

According to Horton ordering of stream begins from the finger-tip tributaries, which do not have their own feeders; rather they are independent in terms of supply of water. Such finger-tip streams are designated as 1st order streams. Two streams of first order, when join together form

2nd order stream. Similarly, two streams of 2nd order meet to make the stream of 3rd order and this process continues till the trunk stream is given the highest order. In Sapahi river basin there are nineteen 1st order stream, four 2nd order and only one 3rd order streams.

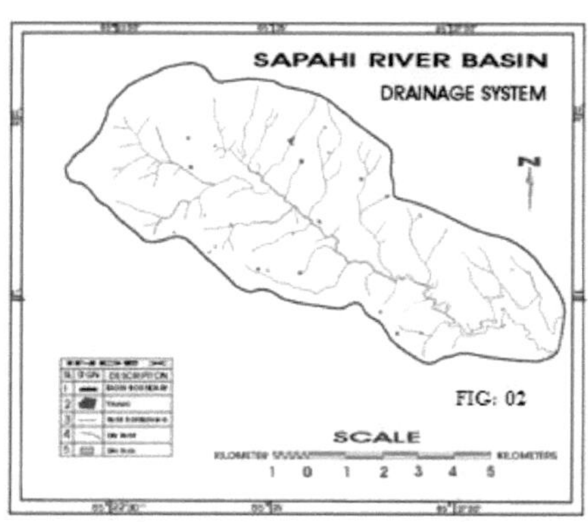

1.3.2 BIFURCATION RATIO (R_b)

Bifurcation ratio, which is related to the branching pattern of the drainage network, is defined as a ratio of the number of streams of a given order to the number of streams of the next higher order and is expressed in terms of the following equation:

$$R_b = N\mu/N\mu+1$$

Where, $N\mu$ = Number of streams of a given order

$N\mu+1$ = Number of streams of the next higher order

Stream Order (μ)	Number of Streams ($N\mu$)	Bifurcation ratio
1	20	4
2	5	2.5
3	2	0

TABLE NO: 01

1.3.3 LAW OF STREAM NUMBERS

The law of stream numbers related with the definite relationship between the orders of the basin and stream numbers.

$$N\mu = R_b^{(k-\mu)}$$

Where $N\mu$ = number of stream segments of a given order

R_b = constant bifurcation ratio

μ = basin order

K = highest order of the basin

= $N_1\text{-}4^{3\text{-}1}$

= $N_1=4^2=16$

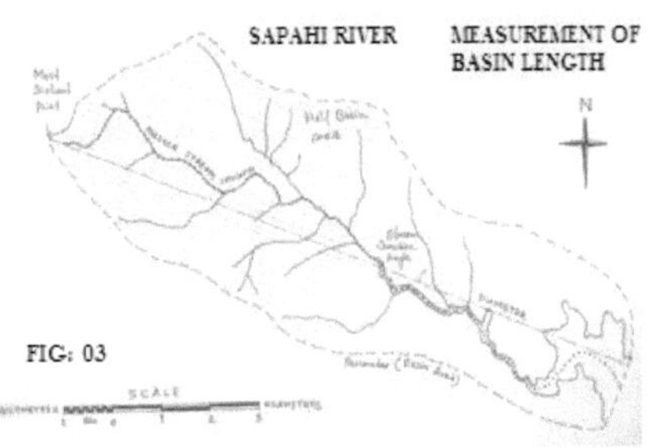

1.3.4 LENGTH RATIO AND LAW OF STREAM LENGTH

The proportion of increase of mean lengths of stream segments of two successive basin orders is defined as length ratio (RI) and is calculated according to the following equation:

$$R_L = L\mu/L\mu_{-1}$$
$$\text{When } L\mu = \Sigma L\mu/N\mu$$

Where, $L\mu$ is the mean length of all stream segments of a given order (μ)

$\Sigma L\mu$ is the sum of lengths of all stream segments of a given order

$N\mu$ is the number of stream segments of a given order.

Stream Order (u)	Number of Streams (Nμ)	Mean length of stream	Lμ
1	20	20	5.17/20=0.2585
2	5	2.5	5.17/5=1.034
3	2	0.67	5.17/2=2.585
	ΣLμ=	5.17	

TABLE NO: 02

1.3.5 **SINUOSITY INDICES** The shape of the open link in terms of geometric structure of observed path (OI) from the expected path – almost line (EI) of a river from the source to the mouth. In fact, no river in practice shows straight path in terms of open link as many causative factors force the drainage line (stream) to deviate from the straight path. The analysis of deviation of the course of drainage line from the straight path says sinuosity.

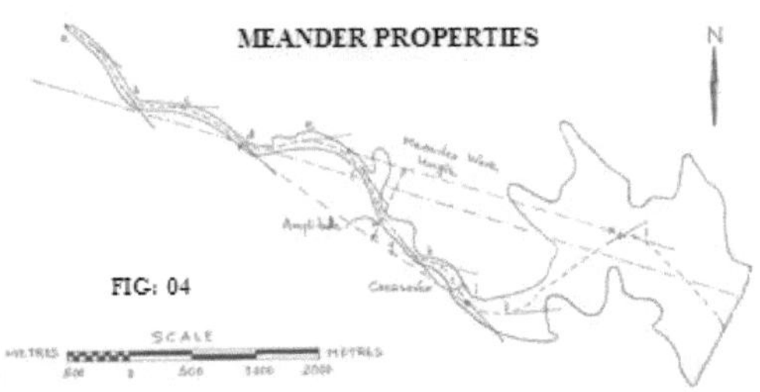

3.6 MEANDER PROPERTIES

The high degree of symmetry of meander length, meander height and form rations is the characteristics of meandering properties which are applied to a sinuous stream that has from place to place along its course or a series of symmetrical arcs, the length of which is related to the width of the stream. J.C. Brice's model of computation of meander properties involves calculation of symmetry of meander length, symmetry of meander height, symmetry of form ratio and repetition of symmetry as follows:

STATION	CURVE VALUE (IN METERS)
A	0.03
B	0.005
C	0.005
D	0.009
E	0.004
F	0.009
G	0.005
H	0.003
I	0.008
J	0.006
K	0.004
L	0.008
M	0.016
N	0.01
O	0.01
P	0.019
Q	0.009
R	0.006
S	0.007
T	0.005
TOTAL	0.11
MEAN DEVIATION OF ARC LENGTH	0.55
MEAN ARC LENGTH	0.0089

Symmetry of (arc) meander length

TABLE NO: 03

7

$$=100 - \frac{100(\text{mean deviation of arc length})}{\text{Mean arc length}} \%$$

$$100 - \frac{100(-0.0055)}{0.0089} \% = 1.66292 \%$$

Generally, the rivers which have these values (symmetry of meander length, meander height, form ratio and repetition of symmetry) above 60% are termed as meandering rivers but some ties terrain and lithological characteristics of rocks restrict the rivers to develop meanders of expected size. According to the above calculation we found that river have very less meandering properties; there is only 1.66292% of river meandering which is not showing any large bending characteristics.

1.3.7 **STREAM JUNCTION ANGLES**

Geometry of angular property of drainage network is related to the junction angle of two stream segments which is defined as the angle projected to the horizontal, between average flow-directions, determined by the ends of the stream segments extending from the junction point to an upstream point at a distance equal to 0.2 times the average length of second order streams (J.K. Lubowe, 1964)

Stream Junction angle:

STREAM JUNCTION	ANGLE (In Degree)	DEVIATION FROM THE BASIN MEAN DIRECTION		SIDE
A	115	135	25	RIGHT
B	105	135	30	RIGHT
C	87	142	55	RIGHT
D	90	150	60	RIGHT
E	90	153	63	RIGHT
F	125	160	35	LEFT
G	259	157	102	LEFT
H	113	147	34	RIGHT
I	110	82	28	RIGHT
J	90	140	50	LEFT
K	100	90	10	RIGHT

TABLE NO: 04

2.2. AREAL ASPECTS OF THE BASIN

Basin area is very important morphometric attribute as it is related to the spatial distribution of a number of significant attributes such as drainage density, stream frequency, drainage texture, slopes, absolute and relative relief, dissection index etc.

2.2.1. GEOMETRY OF BASIN SHAPE

The geometry of basin shape is of paramount significance as it helps in the description and comparison of different forms of the drainage basins and it is also related to the functioning of the units of the basins and its genesis. Thus, various methods have been suggested to calculate the shapes of the basins. On an average three sub-categories of basin shapes have been recognized viz. 1. Circular, 2.Elongated and 3.Indented

A compact shape may be non-elongated, non-indented or slightly indented, whereas a non-compact shape may be elongated, non-elongated, non-indented or highly indented. Different popular methods of computation of basin shape are as follow:

1. **Horton's form factor (F), 1932**

$$F = \frac{A}{L^2}$$

Where, F=form factor indicating elongation of the basin shape
A = Basin Area
L = Basin Length
F = 348/12.5= 27.84 km.

2. **Stoddart's elipticity index (E), 1965**

$$E = \frac{\Pi \pi L^2}{4A}$$

Where E = Elipticity index $\Pi\pi$= 3.14
A = Basin Area,
L = Basin Length

$$\frac{3.14(12.5)^2}{} = \frac{490.625}{} = 2.84$$

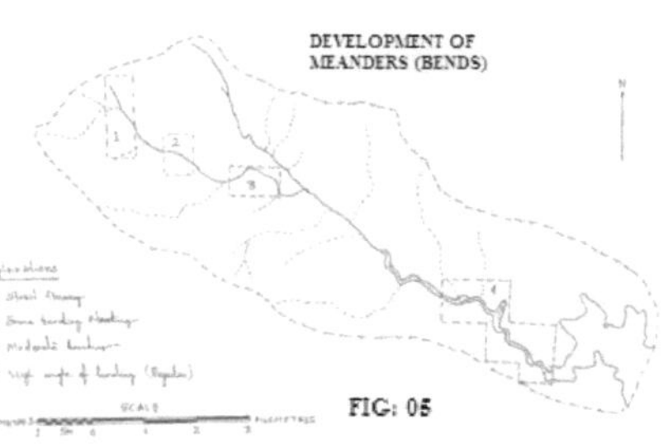

4x348   1392

3. V.C. Miller's Circularity index (C), 1953

$$C = \frac{\text{Area of the Basin}}{\text{Area of the circle with same Perimeter as the Basin (3.75)}}$$

$$C = \frac{4\,\Pi\pi A}{P^2} \quad \frac{4\times3.14\times348}{(3.75)^2} \quad \frac{4370.88}{14.0625}$$

Where P = Basin Perimeter

4. S.A. Schum's Elongation Ratio (R), 1956

$$R = \frac{\text{Diameter of the circle with same area as basin}}{\text{Basin length}}$$

$$R = \frac{3.75}{12.5} = 0.3$$

$$R = \frac{3.14}{4} \, (0.3) \text{ whole sq.} = 0.07 \text{ km}$$

2.3. LAW OF BASIN PERIMETER, BASIN LENGTH AND BASIN AREA

Basin area, Basin perimeter and Channel length are significant morphometric variables wchich determine the shape, size and genetic aspect of drainage basins. Basin perimeter can be directly correlated with the square roots of basin area and increase or decrease in the former indicates increase or decrease in the later.

2.3.1. AREA RATIO (Ra)

Area Ratio denotes proportion of increase of mean basin area between two successive orders and can be calculated by the following equation as suggested by A.N. Strahler (1969):

$$Ra = \frac{A\mu}{A\mu - 1}$$

Where Aμ is mean area of a given order of the basin.

Stream Order (μ)	Area (in sq. km)	Mean Area (in sq. km)	$Ra = \dfrac{A\mu}{A\mu\text{-}1}$
1	230	1.51	2.96
2	40	8.7	1.13
3	78	4.46	1.23

TABLE NO: 05

$$\text{When, } A\mu = \frac{\Sigma A\mu}{N\mu - 1}$$

Where Nμ = Number of all segments of a given order

ΣAμ = Total area of all stream segments of the same order

Since the area becomes cumulative with increasing orders and hence area ratios decrease with increasing order within the basin.

2.3.2. LAW OF BASIN AREA

It may be stated that the basin area becomes cumulative from 1st order to successive higher orders and the trunk system (master stream) of the highest order represents the total area of the whole basin. According to A.N.Strahler, 1969 postulated that, the mean basin areas of successive higher stream orders tend to form geometric series beginning with mean area of the 1st order basin and increasing according to constant area ratio and suggested the following equation basin area:

Stream Order (μ)	Area (in sq. km)	Mean Area (in sq. km)	$Ra = \dfrac{A\mu}{A\mu - 1}$
1	230	1.51	2.96

TABLE NO: 06

$$A\mu = A1\, Ra\,(\mu - 1)$$
$$1.51(2.96)^0 = 1$$

Where A1 is the mean area of the 1st order and

Ra is the constant area ratio

2.3.3. DRAINAGE DENSITY

Drainage density referes to total stream lengths per unit area. It may be defined as a ratio of total length of all stream segments in a given drainage basin to the total area of that basin and thus it can calculate as follows:

$$Dd = \frac{L_k}{A_k} \quad \frac{302.5}{348} \quad = \quad 0.87 \text{ per km}$$

Where Lk   = Total lengths of all stream segments of a the basin
    A_k = Total area of the basin

2.3.4. DRAINAGE TEXTURE

It may be defined as on the basis of stream frequency (number of streams per unit area).

$$Dt = AS = \frac{1}{(t+p)/2}. \quad \text{S. Singh, 1981}$$

Where Dt   = Drainage texture
    As   = Average spacing between two streams

$$t = \frac{(t1+t2)/2}{\sqrt{2}}$$

DRAINAGE STRUCTURE OF SAPAHI RIVER

PHOTO PLATE NO: 04

1. $\dfrac{1/2}{\sqrt{2}} = 0.35$

2. $\dfrac{2/2}{\sqrt{2}} = 0.70$

3. $\dfrac{3/2}{\sqrt{2}} = 1.06$

4. $\dfrac{4/2}{\sqrt{2}} = 1.41$

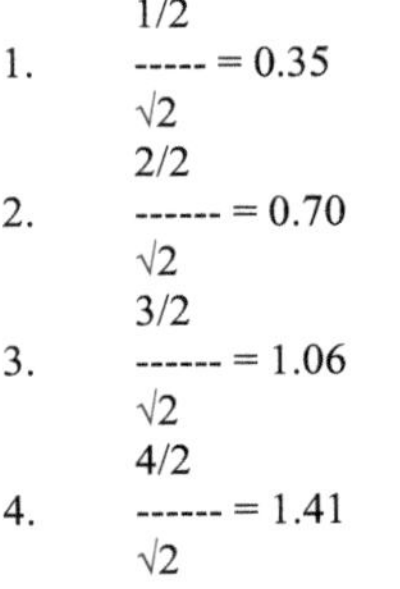

Where t1, t2 = Number of intersections between the stream network and grid square diagonals.
Where p1, p4 = Number of intersections between the stream network and grid square edges.

2.4 RELIEF ASPECTS OF THE BASIN

The relief aspects of the drainage basin are related to the study of three dimensional features of the basins involving area, volume and altitude of vertical dimension of landforms wherein different morphometric methods are used to analyze terrain characteristics, which are the result of basin processes. Thus, this aspect includes the analysis of the relationships between area and altitude (hypsometric analysis), altitude and slope angle (clinographic analysis), average ground slope, relative relief, relief ratio, dissection index, profiles of terrains and the rivers etc.

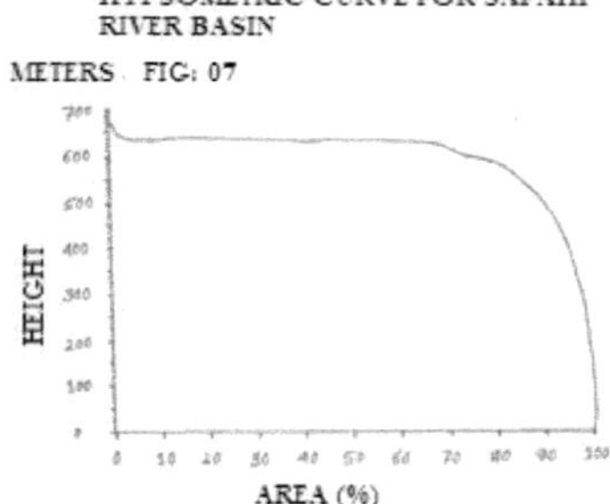

2.4.1 HYPSOMETRIC ANALYSIS

Hypsometric involves the measurement and analysis of relationships between altitude and basin area to understand the degree of dissection and stage of cycle of erosion. Area-height curves, hypsometric curves and percentage hypsometric curves are generally used to show the relationships between altitudes and area of the basin.

The values of Hypsometric Curve of Sapahi river basin

HEIGHT IN METRES	AREA (IN SQ. KM)	PERCENTAGE (%)	% OF QUALITATIVE FREQUENCY
Above 620	57	32.76	0
600-620	85	48.85	81.61
Below 600	32	18.39	100
Total	174	100	

TABLE NO: 07

2.4.2 CLINOGRAPHIC ANALYSIS

Clinographic analysis represent average slopes between successive contours and thus present panoramic view of the terrain. The construction of clinographic curves require data of slope angles between unccessive contours, contour lengths, heights and area between successive contours, for which the following techniques have been adopted.

$$\text{Contours} = \quad \tan\Phi = \frac{CIxL}{A} \qquad \frac{0.02x28.8}{348} = 0.0016^0$$

Where CI   = contour interval
L   = Total length of contour
A   = Total area between contours

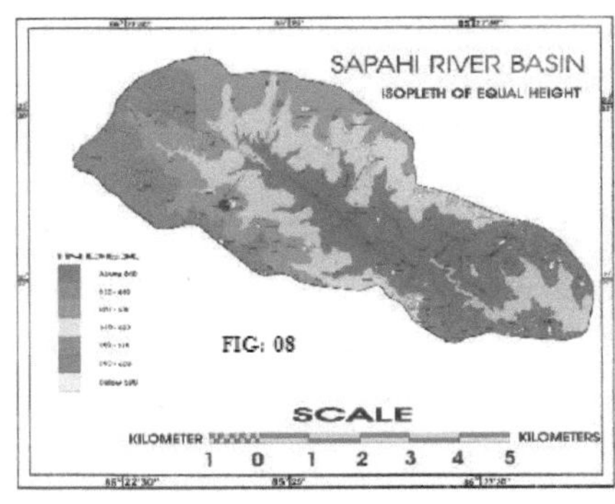

2.4.3 AVERAGE SLOPE

Slopes, defined as angular inclinations of terrain between hill-tops and valley bottoms, resulting from the combination of many causative factors like geological structure, absolute and relative reliefs, climate, vegetation cover, drainage texture and frequency, dissection index etc, are significant morphometric attributes in the study of landform of a drainage basin.

2.5.CHANNEL GEOMETRY OR CROSS-SECTION CHARACTERISTICS

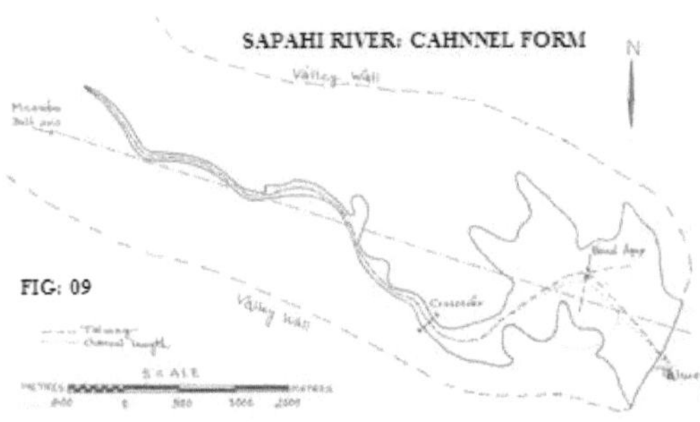

Channel geometry representing the size and shape of cross-sectional and longitudinal channel from includes channel width, channel depth, wetted perimeter, channel slope, channel bends, shape of channel thalwag and their interrelationships. A river channel represents water course of a river confined within the limits of valley walls on both the sides; infect, the river channel presents a three-dimensional form.

2.5.1. CHANNEL LENGTH:

It denotes the distance of channel from source to the mouth of the river falling in that basin and connects all the mid-points across the channel (channel width). The channel length of Sapahi river is about 13.8 km from source in near Jirabar village upto Rukka Dam. The longest tributary is Jirabar which meets with river Sapahi near Jirabar village located in the upper course of the river.

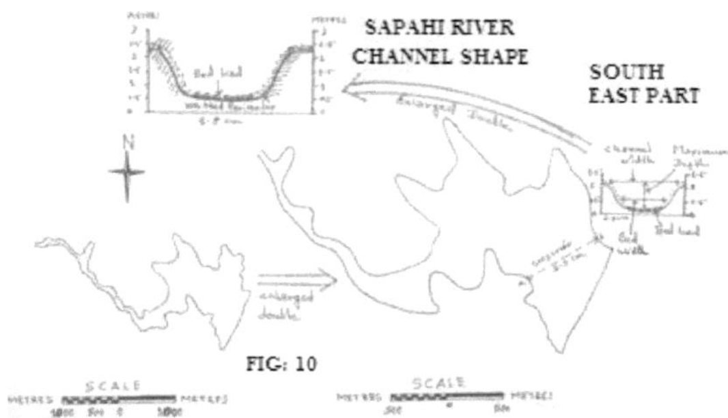

2.5.2. CHANNEL WIDTH:

Channel width (w) at any given point along the course of a river represents straight cross sectional distance of channel representation stage of the river (i.e., level of water). The bank full stage of the river denotes maximum channel width. Thus, channel width considerably changes in the rivers having seasonal regime of rainfall.In dry season we seen in Jirabar river that the channel width was very narrow; water flow was in linear way but in monsoon season it was just opposite. During this season channel becomes wider as 20-30 meter. It occurs due to maximum water runoff/discharge during monsoon rainfall. After this, the maximum channel width is found in Rukka dam where Sapahi meets with Subarnrekha river as 1.4 km. about half of the channel possesses 100 meters. In summer season this width becomes low and DON area uses for agriculture.

2.5.3. CHANNEL DEPTH:

Channel depth is vary from place to place and time to time in the basin, lowest depth is found in upper catchment area (2.3 feet to 3.8 feet); whereas highest depth found in Rukka Dam (5.8feet - 25.6 feet). Again this depth is vary from winter to summer or rainy season.

2.5.4. CHANNEL SLOPE:

The gradient or slope of channel in downward direction is called channel gradient. The whole basin slope / gradient is from west to eastward. Therefore channel slope also follows this direction. Near the west end the channel slope nearly touches to 620 m contour. After following at about 3.8 km the gradient become 600 m. this altitude goes continue till Rukka Dam where Sapahi submerges in Subarnrekha river.

2.5.5. HYDRAULIC GEOMETRY:

The analysis of the relationships among stream discharge, velocity, cahnnel shape, sediment load, channel width, channel depth, channel slope etc. called hydraulic geometry of a river channel. Discharging means outlet system (network) of water mass through forming gullies, channels and main streams flows up to mouth of huge water collection centre, evaporation and sipping toward downward (underground water). Average discharge of river can be calculated as following way:

Discharge Q = w x d x v

Where w = channel width, d = average channel depth, v = average velocity

23/04/09 (summer season):	**Width**	2.7 feet
	Depth	2.3 feet
	Velocity	0.6 feet / sec
	Discharge:	2.7x2.3x0.6 = **3.726 CFS**
09/07/09 (Rainy season):	**Width**	41 feet
	Depth	25.6 feet
	Velocity	1.8 feet / sec
	Discharge:	41x25.6x1.2 = **1,469.44 CFS**

2.5.6. CHANNEL BED TOPOGRAPHY:

The channel bed topography of a river refers to configuration of the river beds in terms of positive and negative features e.g. presence or absence of riffles and pools, sand bars and island, shoals, sand dunes etc. we observed in the basin that in the upper area some small development of gullies are presence. Therefore some amount of sand bars developed. Meanders and valley slides provide sufficient quantity of eroded materials which furthermore works as degradational tools for water. Some pools and riffles also present in the middle of the basin. Big boulders and sand carry out up to Rukka dam, some quantity of alluvial soil also developed through this process.

2.5.7. CHANNEL TYPES:

On the basis of lithological characteristics of the region through which the river has developed its course, the river channels are divided into two broad categories:

1. Bedrock channels and 2. Alluvial channels

In Sapahi river basin bedrock channels are found in the upper catchment area, where gullies and small valley slides provide some metarials for degradational work. In the lower river basin sand deposits are started near Irbba village to Rukka Dam where some quantizes of alluvial soil can be easily identified. Channel pattern is usually associated with alluvial channel. Sapahi river have mostly hard batholiths plateau, that is why geological structure dominantly controls to river valley. Channel pattern can be classified on the basis of sinuosity, channel stability, number of channels (single or multi - thread), slope-discharge relationships, sediment load types etc. 1. Strait channel, 2. Sinuous channel, 3. Meandering channel, 4. Braided channel

Classification of Sapahi River Channel pattern

STATION	CHANNEL TYPE	MORPHOLOGY	SINUSITY	LOAD TYPES	EROSIONAL BEHAVIOUR	DEPOSITIONAL BEHAVIOUR
A	Strait Channel	Single channel with few bedrock	<0.8	Mixture of bedrock	Few Channel widening (DON)	Few Depositional by dryness of water
B	Strait Channel	Single Channel with Pools & Riffles	>0.8	Mixed	Few Channel widening (DON)	Few depositional by dryness of water
C	Sinuous Channel	Single channel with Pools & Riffles Meandering thalweg	>1.05	Mixed	Channel grown up widening	Skew shols
D	Meandering Channel	Single Channel (May be inner point bar channel)	>1.5	Super fine or mixed	Channel increased meandering widening	Point bar formation

TABLE NO: 08   **Note:** Based on self field observation (based on M. Morisawa's classification)

CONCLUSIONS

On the basis of above discussion the following conclusion may be conclude:

1. Sapahi river basin is completely seasonable, hydraulic geometry is changing from time to time. For instance, in summer season **(23/04/09)** discharge were about **3.726 CFS**, whereas in rainy season **(09/07/09)** it was **1,469.44 CFS**.

2. As per Stream ordering method of Strahler Sapahi river possesses up to third order of stream ordering. It has drainage density between **0.2 – 0.3** and drainage shape is of **0.4**.

3. Altitude study of the basin is showing that there is generally $2^012'$ to $4^030'$ of gradient, whole river basin have sloping from north-west to south-eastward.

4. The highest point is of **617 meters** which is located near **Haldama village** in north-west portion of the river basin.

5. In the river valley three types of channels are found, viz, **strait, sinuous** and **meandering**.

6. Meandering property includes bending, having mean length of **0.0089**.

7. **57 sq. km** area lies under 620 meters height, **85 sq. km** under 600-620 meters and **32 sq.** km lies within 600 meters of contours.

8. Sapahi river submerse their water in Rukka Dam after successfully flowing of **13.5 km basin length**.

9. Whole river basin contains some **peneplain property** which shows their degradation history of **Ranchi plateau**.

10. Earlier establish hypothesis that '**Discharge of Sapahi river basin differs in different seasons**', is proved through conclusion number one.

11. **Basin morphometry depends on the configuration of surface and runoff characteristics** which confirmed through above different morphological analysis.

12. Thus, earlier said that, Sapahi river runs over Ranchi plateau which developed over hard batholithical surface, it gives a permanent shape for river drainage network; that's mean river cannot change their path such as in alluvial deposited river (Sapt Koshi or Gandak river). River meandering occurred very less, the established hypothesis that, **Degradational behavior of Sapahi river basin is based on the Lithological characteristics** is proved.

REFERENCE:

- Marrisawa, M., 1968, Streams their dynamics and Morphology; Mc. Graw-Hill Book Co., Newyork, pp- 18-40, 46-76, 82-110.

- Mukhopadya, S.C., 1982, The Tista Basin; A study in Fluvial Geomorphology, K.P. Bagchi and Co., Kol, 308 p.

- Mukhopadya, S.C., 1980, Geomorphology of the Subarnarekha Basin, The Univ. of Burdwan, pp- 172-191.

- Sarkar, G. & Gupta, S., 2009, Fluvio-Geomorphological characteristics of Downstream of R. Panchonoi. Ind. Jour. Of Landscape system and ecological studies, Institute of Landscape, Vol-32, No.1, pp-9-14.Kattelmann.; publ. no. 201, 1991, Hydrologic regime of Sapt Kosi Basin, Nepal.

- Schumm, S.A. and H.R. Khan, 1972; Experimental study of channel patterns. Geological Society of America, Bulletin, 83, 1755-70.

- Wadia, D.N. (1975), Geology of India, Tata Mc. Graw-Hill, New Delhi, pp-10-250.

- Mukhopadya, S.C., 1980, Geomorphology of the Subarnarekha Basin, The Univ. of Burdwan, pp- 172-191.

- Mukhopadya, S.C., 1982, The Tista Basin; A study in Fluvial Geomorphology, K.P. Bagchi and Co., Kol, 308 p.

- Geographical perspective, A journal of the AGBJ, 2008 Vol-9.

- District Census Handbook, 1981, Primary Census Abstract, District –Ranchi, Series-4, Part XIII-A& B.

- Mahato, R.K., 2009, Morphometric analysis of Sapahi river (Unpublished Master thesis); pp- 35-48.

- Talash, (2 Monthly magazine), Series 4, 1 Nov, 2008.

- Ranchi District Gazetteer, 1954, pp-25-27.

- www.jharkhandgov.com

- www.encyclopedea.com